W9-DDP-819

FAST FORWARD

Concept Cars & Prototypes of the Past

Publications International, Ltd.

Louis Weber, CEO
Publications International, Ltd.
8140 Lehigh Avenue
Morton Grove, IL 60053

ISBN: 978-1-64030-650-9

Manufactured in China.

8 7 6 5 4 3 2 1

PHOTOGRAPHY:
The editors would like to thank the following people for supplying the photography that made this book possible. They are listed below, along with the page numbers of their photos.

Thomas Glatch: 28-29; Vince Manocchi: 82-83; Al Rogers: 4-5, 26-27, 32-33, 68-69; Alex Steinberg: 70-71.

OWNERS:
Special thanks to the owners of the cars featured in this book for the cooperation.

Joseph Bortz: 26-27, 28-29, 32-33, 82-83; Detroit Historical Society: 68-69; Thomas G. Maruska: 70-71; Sloan Museum: 4-5.

Special thanks to following manufacturers who supplied imagery.

Aston Martin Lagonda Limited, BMW Group, Daimler AG, Fiat Chrysler Automobiles, Ford Motor Company, General Motors Company, Hyundai Motor Company, Jaguar Land Rover Limited, Automobili Lamborghini S.p.A, Mazda Motor Corporation, Nissan Motor Corporation, Rolls-Royce Motor Cars Limited, Tesla Inc, Volkswagen AG

BUICK XP-300

The 1951 Buick XP-300 was a project of Charles Chayne, Buick's chief of engineering 1936-1951. While most concept cars were more show than go, the XP-300 was powered by a 335-horsepower supercharged

M112·1
MFR

Buick

M112·1
MFR

1951
Le Sabre
AE·4940
USA

GM LESABRE

The 1951 GM LeSabre was developed at the same time as the Buick XP-300. While the XP-300 was the responsibility of Buick engineer, Charles Chayne, the LeSabre was a project of legendary General Motors styling chief, Harley Earl. Both dream cars shared a 335-horsepower supercharged V-8.

1953

BUICK WILDCAT

The General Motors Motorama show was a lavish extravaganza that toured the nation to showcase GM production cars and concept cars. A star of the 1953 Motorama was the Buick Wildcat. Many design features of the fiberglass dream car found their way to Buick production cars.

CHRYSLER D'ELEGANCE

The 1953 Chrysler D'Elegance was designed Chrysler stylist Virgil Exner working with Italian coachbuilder Ghia. The show car was mounted on a shortened Chrysler New Yorker chassis. The design was later modified to create the Volkswagen Karmann-Ghia.

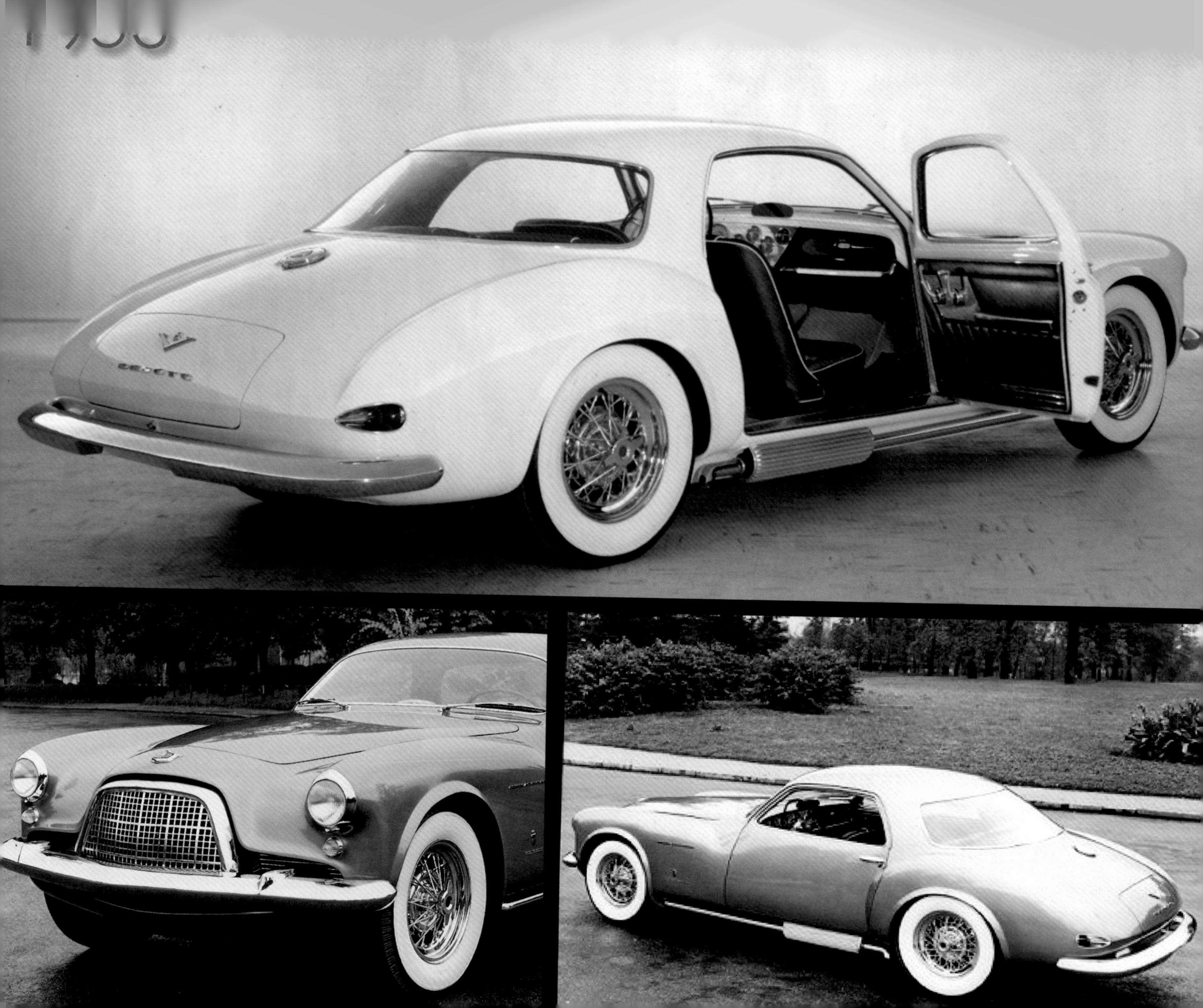

DESOTO ADVENTURER

Virgil Exner designed a series of dream cars for Chrysler Corporation in the Fifties that were fabricated by Italian coachbuilder Ghia. The 1953 DeSoto Adventurer was powered by a stock DeSoto 276-cid FireDome "hemi" V-8 with 170 horsepower. The Adventurer name was used on DeSoto's flashy, high-performance coupes and convertibles from 1956 to 1959.

LINCOLN XL-500

Fifties stylists seemed to believe that bubble-top cars were the future and many show cars had clear roofs. Concept cars were built to test styling ideas and many, such as the 1953 Lincoln XL-500, lacked an engine and other mechanical elements. The XL-500 had to be pushed on stage. Such cars were known as "pushmobiles".

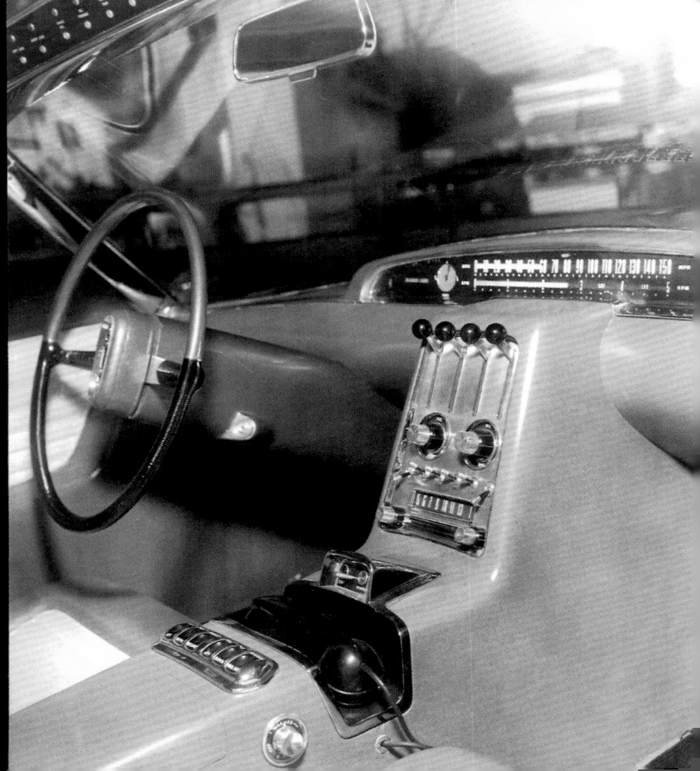

1954

BUICK WILDCAT II

The 1954 Buick Wildcat II, the second of Buick's Wildcat show cars, was a sports car that was inspired by the Chevrolet Corvette concept car. The Wildcat II was powered by a modified Buick V-8 with four carburetors that was rated at 220 horsepower.

CADILLAC EL CAMINO

Cadillac built several two-seater show cars in the Fifties, although a production two-seater Caddy didn't appear until the 1987 Allanté. The Cadillac El Camino was a fiberglass coupe that appeared in the 1954 Motorama. The El Camino name would later appear on Chevrolet's car-based pickup.

1954

1954

FORD FX ATMOS

The 1954 FX Atmos was Ford's first "spaceship" concept car. The driver sat in a center seat with two handgrips instead of a steering wheel. If the inoperable fiberglass dream car had been equipped with an engine, it would have been mounted in the rear.

GM FIREBIRD I

The 1954 Firebird I was the first of General Motors' four gas-turbine powered dream cars. The jet-like engine developed 400 horsepower. The Firebird would later lend its name to Pontiac's ponycar.

1954

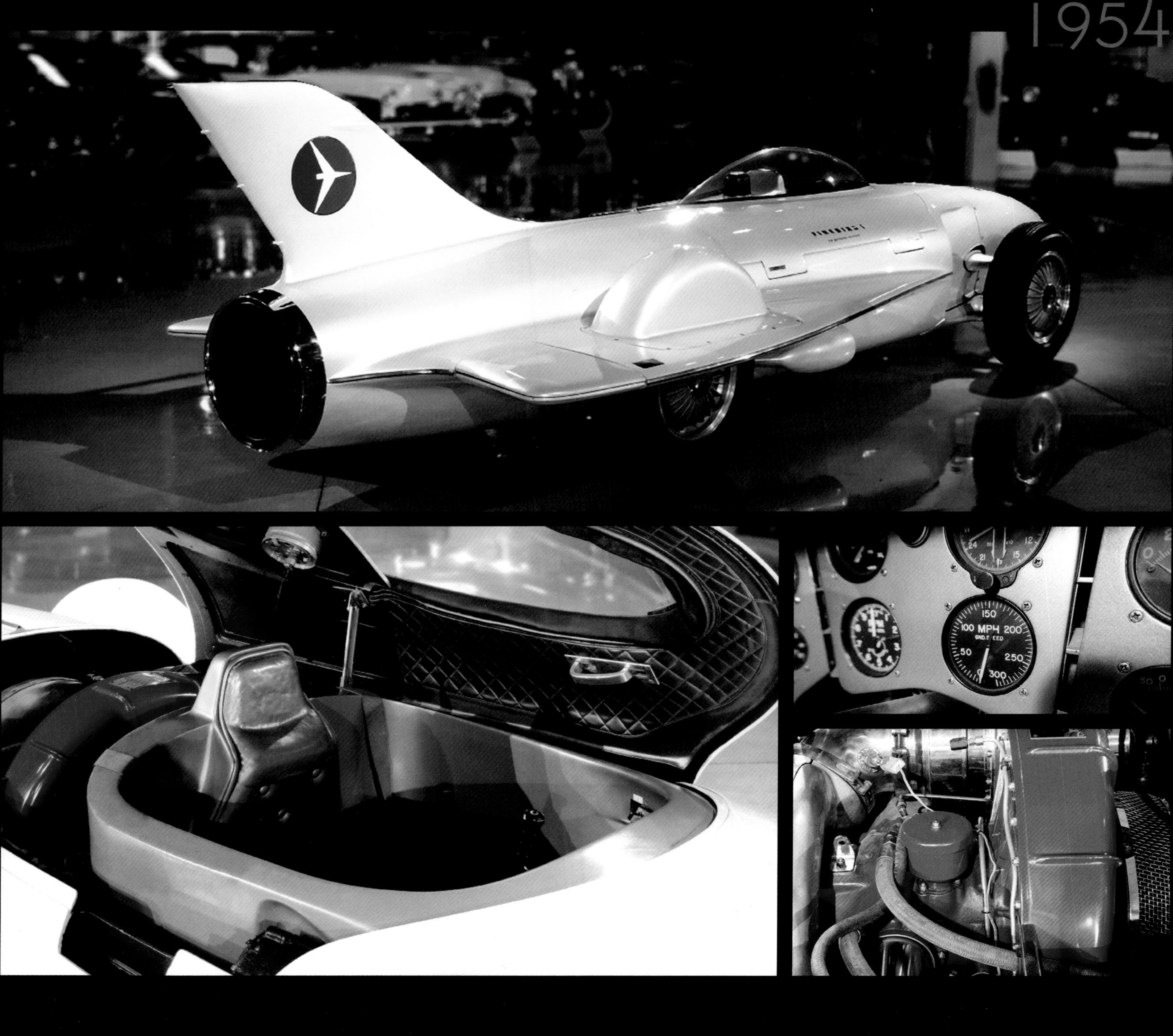

1954

OLDSMOBILE CUTLASS

The 1954 Oldsmobile Cutlass was a fastback coupe with a distinctive louvered rear window. Cutlass likely drew its name from the Chance-Vought Cutlass fighter jet. The Cutlass name appeared on a compact coupe in 1961 and was a common Oldsmobile model name thereafter.

PONTIAC BONNEVILLE SPECIAL

Following the success of the Chevrolet Corvette concept car, other General Motors divisions wanted to get in on the act. The 1954 Pontiac Bonneville was a futuristic sports car that took its name from the Bonneville Salt Flats in Utah. Powering the fiberglass coupe was a modified version of Pontiac's aging straight eight. Pontiac introduced a modern V-8 the following year.

BONNEVILLE-SPECIAL
000
UTAH 1954

BONNEVILLE-SPECIAL

1955

CHEVROLET BISCAYNE

The 1955 Chevrolet Biscayne was a hardtop sedan with no center pillars— the front and rear doors latched on each other. Another unusual feature was a windshield that not only curved around the sides, but also extended into the roof.

FORD MYSTERE

The 1955 Ford Mystere had several styling details, such as the bodyside molding dip and the rear fins, that would show up on the 1957 Ford Fairlane. Rather than tip off the competition too soon, the Mystere was kept off the auto-show circuit until late 1955.

GM LASALLE II

LaSalle was a companion brand built by Cadillac from 1927 to 1941 and for several decades General Motors hinted that the make might be revived. Styling chief, Harley Earl, got his start at GM by designing the first LaSalle. For the 1955 GM Motorama, Earl built two LaSalle II concept cars to showcase a V-6 engine that the corporation was developing. Besides this roadster, there was also a LaSalle II hardtop sedan.

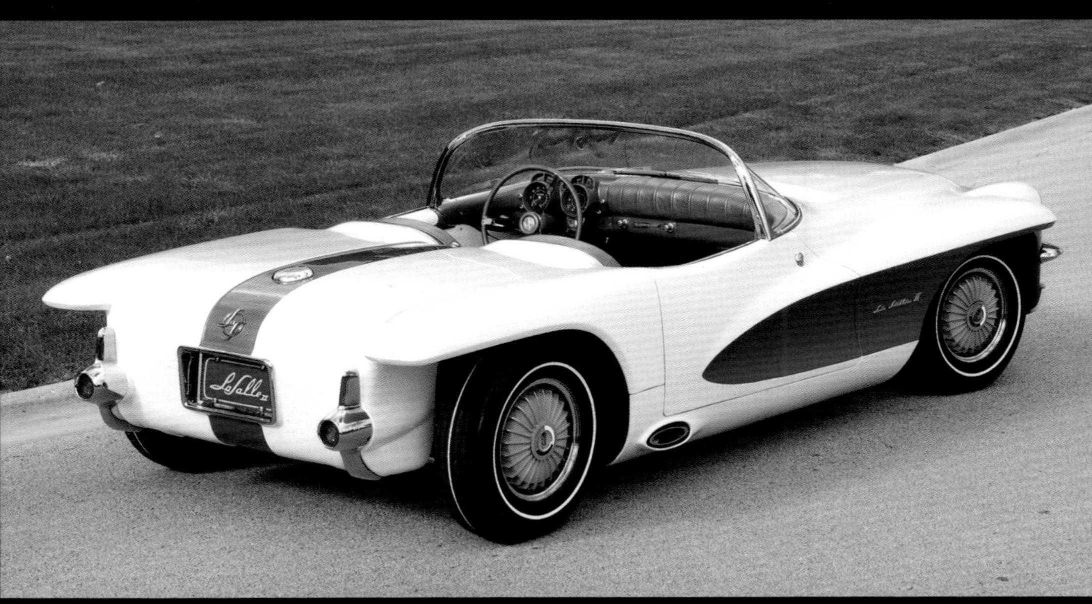

La Salle II

1955

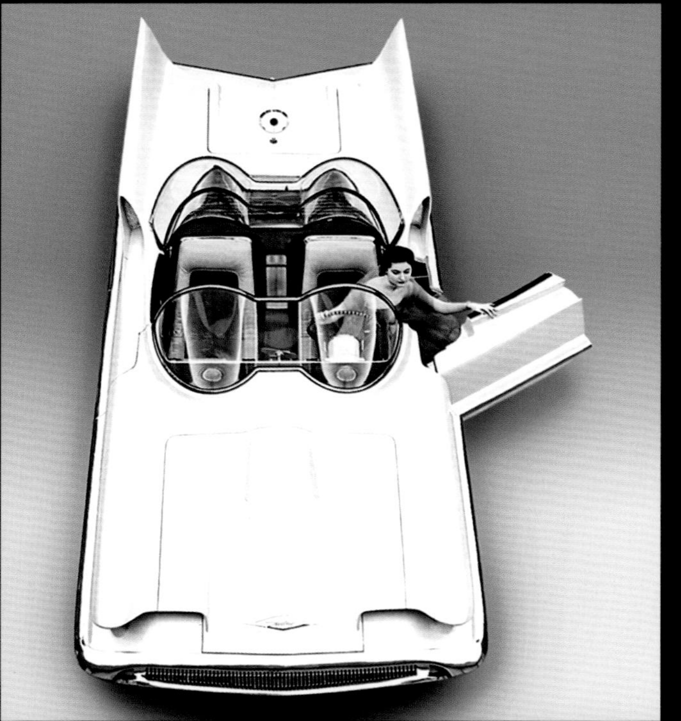

LINCOLN FUTURA

The 1955 Lincoln Futura was one of Ford Motor Company's most successful show cars. Unlike many concept cars, the Futura was fully functional. In 1965, California car customizer George Barris received a request to construct a car for the *Batman* TV show. With some sheet-metal modification, along with the addition of "bat" equipment, the Futura was reborn as the Batmobile.

BUICK CENTURION

The 1956 Buick Centurion was a jetfighter-inspired concept car with a Plexiglas canopy. Rearview mirrors were replaced by a closed circuit TV system with a screen on the instrument panel. A Buick V-8 was tuned to produce 325 horsepower—70 more than the standard Buick engine of 1956.

1956

1956

GM FIREBIRD II

The second gas-turbine powered General Motors Firebird concept car had seats for four and was billed as the "Family Sedan of the Future". The 1956 GM Firebird II had an electronic guidance system that would follow wires buried in the roadways of the future—an early concept for a self-driving car.

MERCURY XM-TURNPIKE CRUISER

The 1956 Mercury XM-Turnpike Cruiser concept car was styled at the same time as the production 1957-58 Mercury Turnpike Cruiser. The dream car was tapped for ideas that turned up on production Mercs, such as: concave fins ending in wedge-shaped taillights; a concave grille; and forward-thrusting hoods over the headlights. The Plexiglas roof panels, which lifted up gullwing-style when the doors were opened, were not included on production cars.

GENERAL
G

OLDSMOBILE GOLDEN ROCKET

The 1956 Oldsmobile Golden Rocket was a futuristic two-seater with a split rear window similar to the 1963 Chevrolet Corvette Stingray. The roof panels were automatically lifted when a door was opened. Oldsmobile first used the Rocket name in 1949 for their influential overhead-

PLYMOUTH PLAINSMAN

Not all concept cars were sports cars. This 1955 Plymouth Plainsman was Virgil Exner's take on the practical station wagon. The Plainsman had a Western theme with unborn calfskin upholstery and longhorn emblems. A power-folding third-row seat and power tailgate are two Plainsman features that are popular on today's SUVs.

1956

SURF CLUB
APARTMENTS
9133
MEMBERS ONLY

here's where the new ideas start...
PONTIAC
Club de Mer
More than a glamorous dream car, the unique Pontiac Club de Mer is an exciting experimental laboratory on wheels. Many of its engineering and design innovations you might see on the Pontiacs of the future. Such things as the Club de Mer's anodized brushed aluminum body, extreme road-hugging lowness (a scant 38 inches), jet-like stabilizing fin and bubble windshields are indications of things to come. The Club de Mer, not planned for production, is powered by a 300-horsepower Strato-Streak V-8 with dual four-barrel carburetors.

PONTIAC CLUB DE MER

The 1956 Pontiac Club de Mer was the most radical of the 1956 General Motors concept cars. The aluminum-bodied roadster was only 38 inches high and had twin Plexiglas bubbles instead of a windshield. The Club de Mer had a steel tube chassis to which a modified stock Pontiac front suspension was attached.

FORD LA GALAXIE

The 1958 Ford La Galaxie was a fiberglass concept car with many far-out ideas, although none were functional. Power was to be provided by a nuclear engine and a screen would have informed the driver of his proximity to other vehicles. Predicting today's autonomous emergency braking systems, the La Galaxie's brakes would have been applied automatically if the car got too close to another object.

FORD

1958

SIMCA FULGAR

The United States didn't have a monopoly on dream cars. The French Simca Fulgar was a car of the future with nuclear power, voice control, and radar guidance. Above a set speed, two of Fulgar's four wheels were to retract and the car would be balanced by gyroscopes. None of these futuristic features functioned, but Simca was looking to the future.

CADILLAC CYCLONE

The 1959 Cadillac Cyclone was one of the last dream cars commissioned by General Motors styling chief Harley Earl. The front nosecones contained a radar system that acted as a proximity warning system. A sensor in the car could follow wires that GM predicted would be buried in the highways of the future. This early version of a self-driving car was said to have worked well on the General Motors proving grounds. In common with 1959 production Cadillacs, the Cyclone had outlandish fins. However, these were later cut down to a more modest height.

CHEVROLET CORVETTE STINGRAY

General Motors' new design chief, Bill Mitchell, got around the corporate racing ban by building the 1959 Chevrolet Corvette Stingray race-car with his own funds. The chassis was from an aborted Corvette SS Le Mans racer. The body was based on a "Q-Corvette" styling study for a cancelled redesign of Corvette for 1960. The Stingray raced successfully and won the 1960 Sports Car Club of America C-Modified class. When Corvette was finally redesigned for 1963, it was influenced by the Stingray.

1959

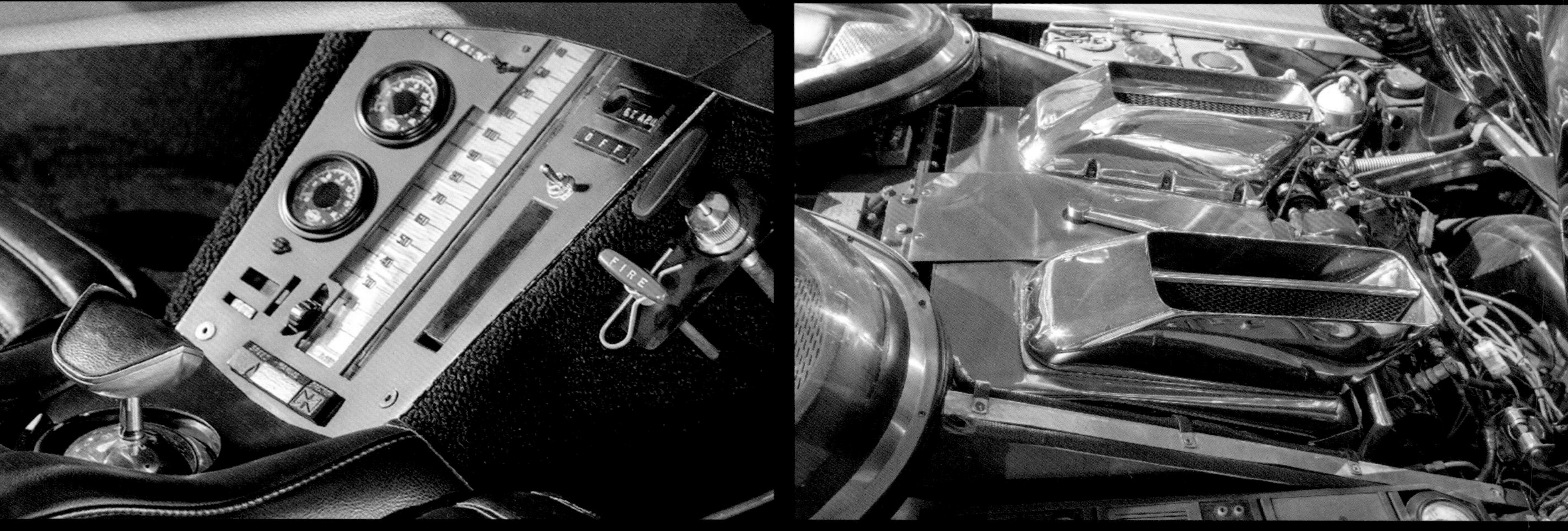

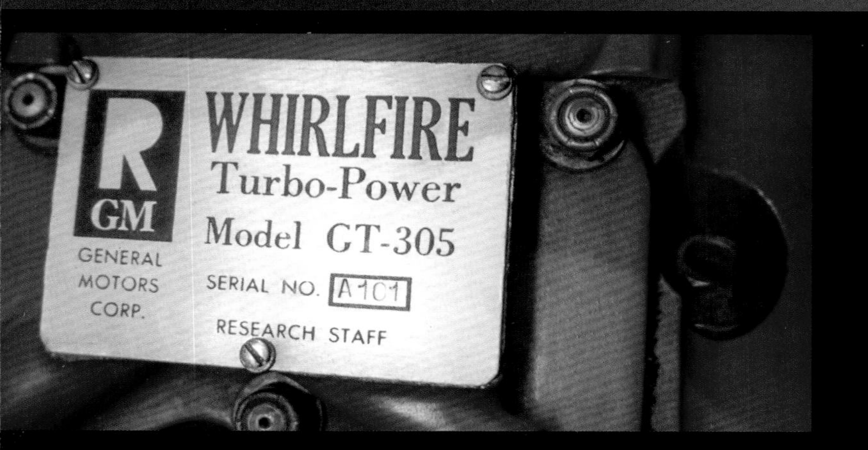

GM FIREBIRD III

The third General Motors Firebird gas-turbine-powered concept car took the Fifties tailfin fad to the limit. The fins on a Nike two-stage missile inspired the design, which had a total of seven tailfins. The 1959 Firebird III was controlled by a joystick that allowed the driver to go, stop, and steer with a single lever.

PLYMOUTH XNR

Chrysler Corporation styling chief Virgil Exner used the compact Plymouth Valiant as the base for a sports car prototype. The Valiant's 225-cid Slant Six was modified to develop 250 horsepower. The car was said to have lapped the Chrysler test track at more than 150 mph. Exner hoped there would be a production version in Plymouth showrooms, but the project ended with the prototype.

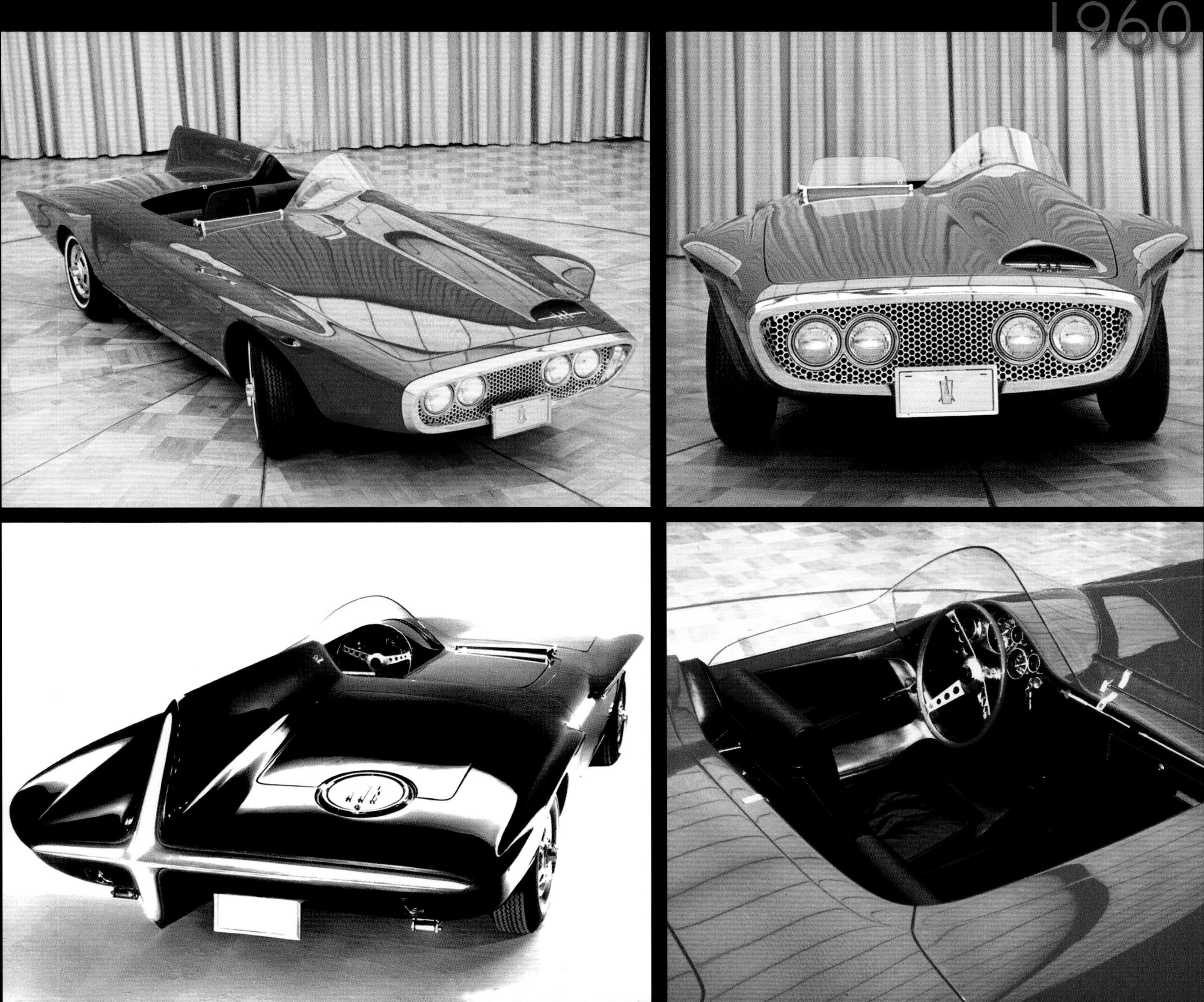

1961

CHEVROLET MAKO SHARK CORVETTE

A shark caught by General Motors design chief William Mitchell inspired the styling of the 1961 Chevrolet Mako Shark Corvette show car. Corvette was overdue for a redesign and the Mako Shark helped generate interest until a redesigned 'Vette arrived for 1963. The new Corvette showed the influence of both the Mako Shark and Mitchell's 1959 Corvette Stingray racecar.

CHRYSLER TURBOFLITE

Chrysler was serious about bringing the jet turbine engine into the family car. The 1961 Chrysler Turboflite concept car put the company's latest turbine engine in a car with jet-fighter-inspired styling. The Turboflite was topped with a canopy that lifted automatically when a door was opened.

1961

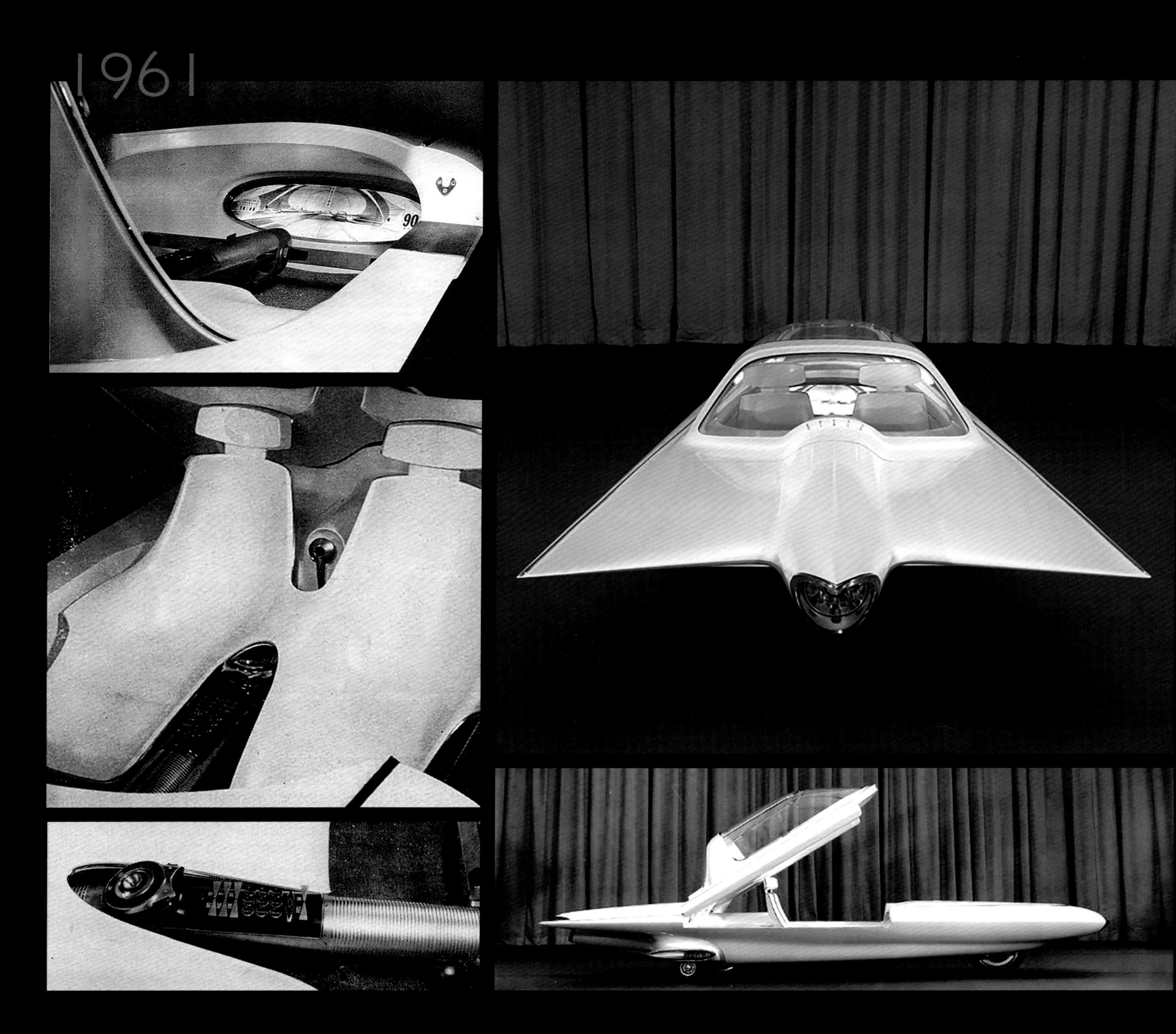

FORD GYRON

Prolific designer Alex Tremulis was perhaps most famous for his work on the 1948 Tucker. Tremulis later worked at Ford where he was chief of advanced styling and designed the 1961 Ford Gyron concept car. The aerodynamic Gyron was intended to drive on two wheels with a gyroscope providing balance, but a gyroscope was never installed in the show car and a pair of outrigger wheels kept the Gyron upright.

CHEVROLET CORVAIR SUPER SPYDER

The Chevrolet Corvair was conceived as an economy car, but found a niche as a sporty compact—especially the turbocharged Spyder models. The two-passenger 1963 Chevrolet Corvair Super Spyder concept car was built on a shortened Corvair platform with an impractical, but racy, windshield and six chrome exhaust pipes.

1963

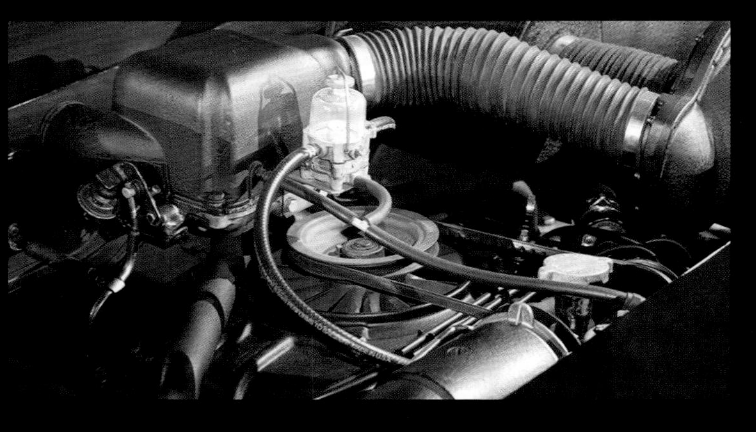

1963

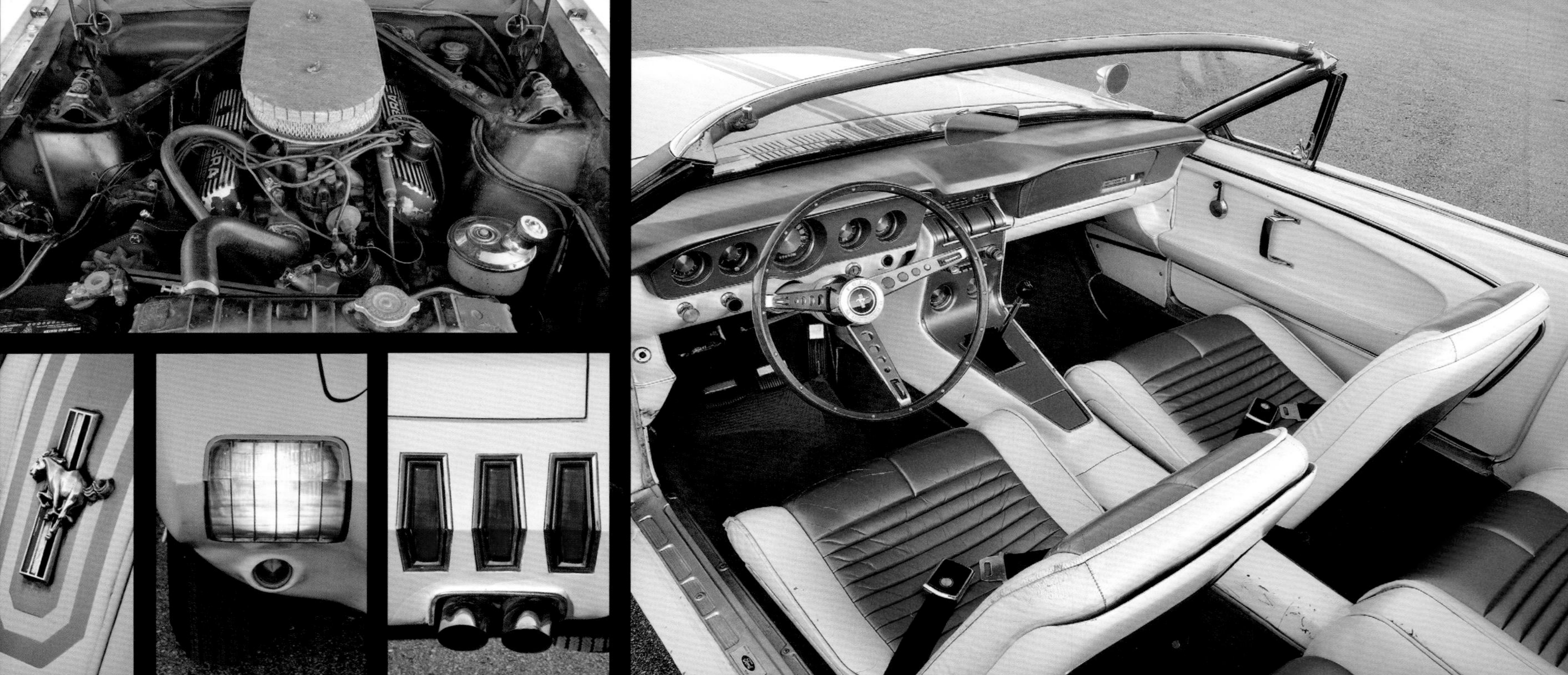

FORD MUSTANG II

The first Ford Mustang I concept car was a two-passenger midengined sports car, but Ford decided a more conventional four-seater would sell better. The 1963 Ford Mustang II show car (not to be confused with the 1974-78 production Mustang) was built as a link between the Mustang I and the upcoming production car. The Mustang II used a production Mustang body with a customized nose and tail that echoed the original Mustang I.

FORD THUNDERBIRD ITALIEN

The 1963 Ford Thunderbird Italien was a production Thunderbird modified with a fiberglass fastback roof and other unique features. The Italien toured the country as part of Ford's Custom Car Caravan and appeared at the 1964 New York World's Fair in Ford's Cavalcade of Custom Cars. The Italien was featured in several magazines at the time.

1963
ITALIEN
Thunderbird
FoMoCo

1964

RIGHT
AM 5 6 7 8 10 15 KC
FM 89 92 94 98 102 106 MC
50 60 70 80 90
COMFORT CONTROL

FORD AURORA

The 1964 Ford Aurora was a station wagon of the future that debuted at the 1964 New York World's Fair. The third-row seat was accessed by a two-piece rear tailgate with a lower section that doubled as stairs. The lounge-like center section featured an L-shaped middle seat and a combination refrigerator/oven unit. The instrument panel had an early version of GPS with a rolling map.

FORD GAS TURBINE

With America's interstate system expanding in the Sixties, the trucks of the future would have to be bigger and faster. The 1964 Ford Gas Turbine prototype (commonly known as "Big Red") was 13 feet tall and 96 feet long. A 600-horsepower gas-turbine engine could propel the truck 70 mph when fully loaded or 78 mph partially loaded. Big Red crossed the nation several times on a promotional tour and drivers reported

1964
FORD MOTOR COMPANY
new horizons in superhighway transport

1964

GM BISON

The 1964 General Motors Bison was GM's vision of the truck of the future. Displayed at the 1964 New York World's Fair, Bison placed the driver in a pod in front of the wheels, while the twin turbine engines were concealed in a pod over the wheels. Unlike Ford's Gas Turbine truck that was driven across the country, the Bison was a nonoperative concept.

GM FIREBIRD IV

The 1964 General Motors Firebird IV was the final concept car of the Firebird series. While the other Firebirds had functioning gas-turbine engines, the Firebird IV was a "pushmobile" that only claimed to be gas-turbine powered. Also nonfunctioning was a computer guidance system that would have relied on wires buried in the highways of the future. In 1969, a slightly revised Firebird IV returned as the Buick Century Cruiser.

1964

GM X STILETTO

The cleanly styled 1964 GM-X Stiletto lacked conventional doors. The roof and tail section lifted up and passengers entered through a rear opening. General Motors stylists liked aviation themes. That was most obvious on the instrument panel that had more switches and gauges than a 747. A revised Stiletto later returned as the Pontiac Cirrus in 1970.

PONTIAC BANSHEE

While chief engineer for Pontiac, John Z. DeLorean pushed for a two-seat sports car, and Banshee coupe and convertible concept cars were built. Engines included an overhead-cam six and V-8s. If it had reached production, the Banshee would probably have been priced below Chevrolet's Corvette.

1964

Banshee
STEEL BELTED RADIAL

BANSHEE
DEC 94
STEEL BELTED RADIAL
UNIROYAL

Banshee

PONTIAC
MOTOR DIVISION

CADILLAC XP-840 ELDORADO

In the Sixties Cadillac considered building either a V-12 or V-16 engine. In 1965, Cadillac styled a full-sized mockup for a two-seat V16 fastback coupe. The Cadillac XP-840 Eldorado had no rear window; instead, a narrow slit was cut in the roof for a rear-facing TV camera. Cadillac dropped the V-12 and V-16 engine idea and the XP-840 never got beyond the styling study stage.

DODGE CHARGER III

The 1968 Dodge Charger III concept car was far different from the production Dodge Charger. Instead of a muscle car, the Charger III seemed to be aimed at the Chevrolet Corvette market. Like many concept cars of the Fifties and Sixties, the Charger had a canopy instead of conventional doors. Braking was aided by three flaps in the rear that could deploy as air foils.

GOODYEAR
Charger III
GOODYEAR

GOODYEAR
Charger III
GOODYEAR

1969

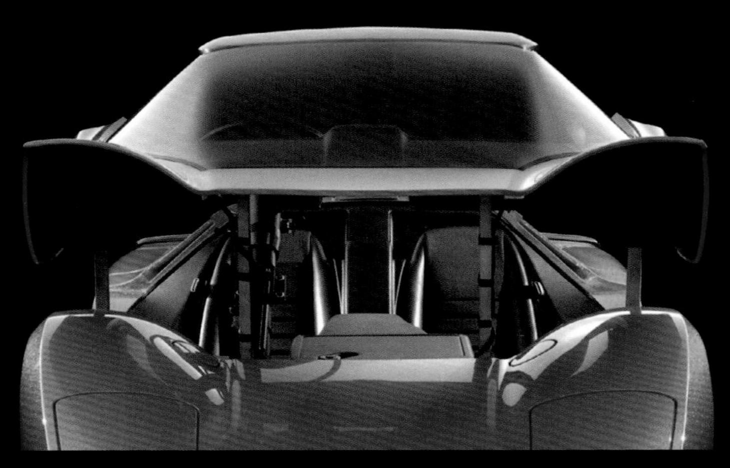

HOLDEN HURRICANE

Holden, General Motors' Australian division, shocked the world with its first concept car in 1969. The Holden Hurricane was a sports car with a new 4.2-liter V-8 mounted midship. Advanced features included on-board navigation and a camera instead of a rear window. Although the Hurricane never entered production, its V-8 engine was built by Holden for many years.

CHRYSLER CORDOBA DEL ORO

Elwood Engel designed the 1961 Lincoln Continental before coming to Chrysler. Similar to the Continental, the 1970 Chrysler Cordoba Del Oro was a large car with clean lines. A cantilevered roof allowed extremely thin windshield pillars. The Cordoba name would later be used on Chrysler's personal luxury coupe.

CHRYSLER
Plymouth

LANCIA STRATOS HF ZERO

Italian coachbuilder Bertone took the wedge-shape, midengined sports car design to the limit with the 1970 Lancia Stratos HF Zero concept car. In spite of its extreme design, the Stratos was drivable with power provided by a Lancia V-4 engine. Passengers entered the Stratos by lifting the windshield. Lancia later built a more conventional production Stratos that did well in the World Rally Championship series.

CHEVROLET AEROVETTE

Corvette started experimenting with midengined Corvettes in 1969. In 1973, Chevrolet displayed a midengined Aerovette concept car on the auto show circuit that was close to its planned 1980 production derivation. But a mid-engine Corvette was not to be—at least not in the Eighties. Chevrolet decided to continue with the conventional front-engine Corvette.

1973

BUICK WILDCAT
BUICK

1985

BUICK WILDCAT

The 1985 Buick Wildcat used a version Buick's V-6 with modifications by racing specialist McLaren. The engine was mid mounted and drove all four wheels. Entry required pushing a solenoid, which raised the canopy.

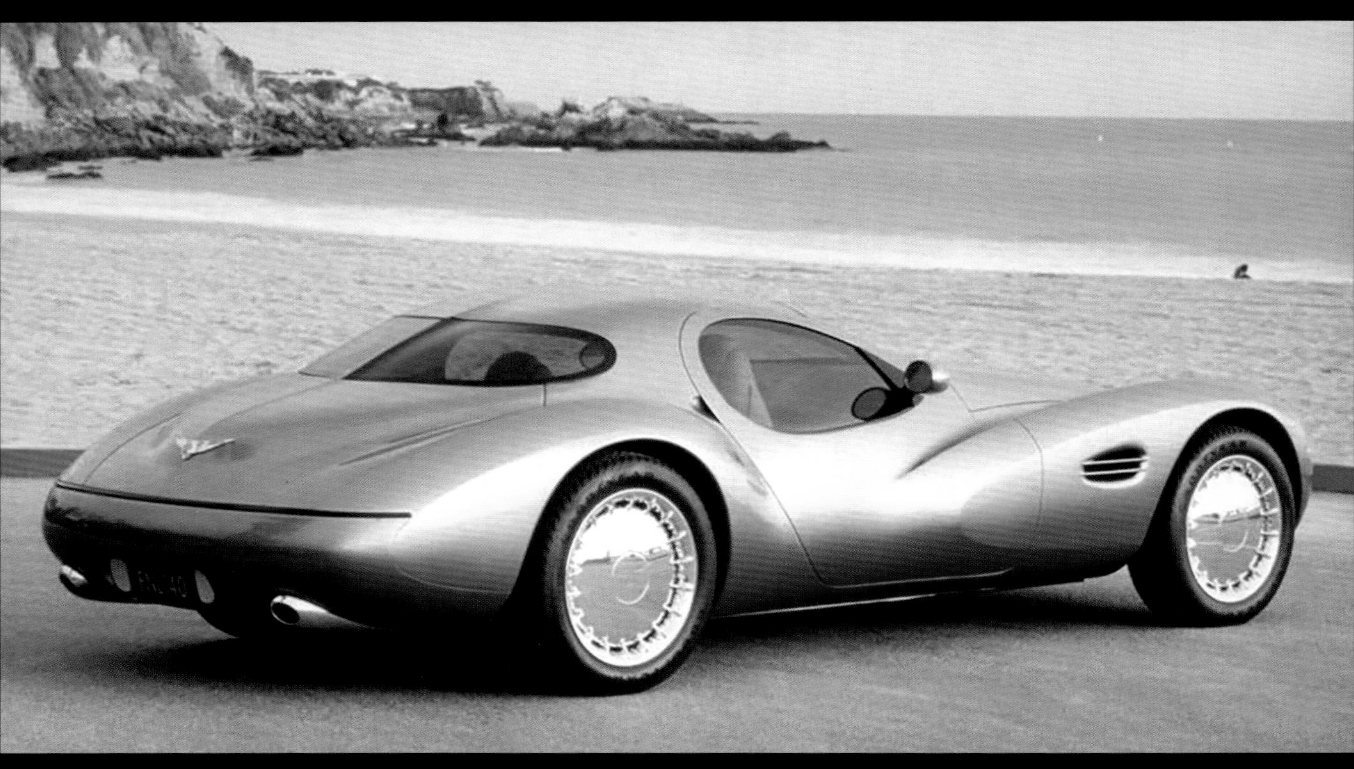

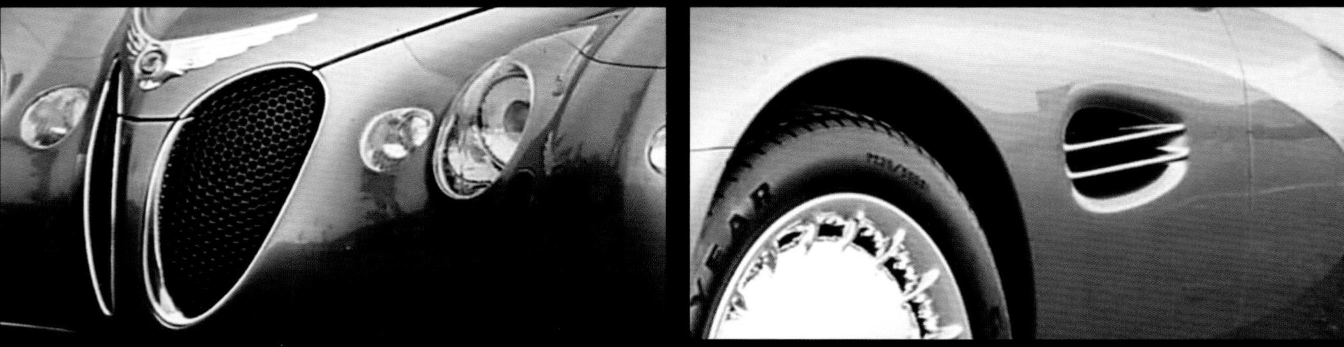

CHRYSLER ATLANTIC

The 1995 Chrysler Atlantic was inspired by pre-World War II sport coupes, including the French Bugatti Type 57 Atlantic. Like the Bugatti, the Chrysler was powered by a dual-overhead-cam straight-eight engine. Chrysler combined two Dodge Neon four cylinders to create an engine layout the corporation hadn't used since the Fifties.

1995

FORD 49

The 1949 Ford's styling and engineering were a huge step forward and a turning point for the company's fortunes. The 2001 Ford 49 concept harked back to the 1949 model in styling, but with a modern flair. Under the hood was a dressed-up version of a 3.9-liter Thunderbird V-8.

2001

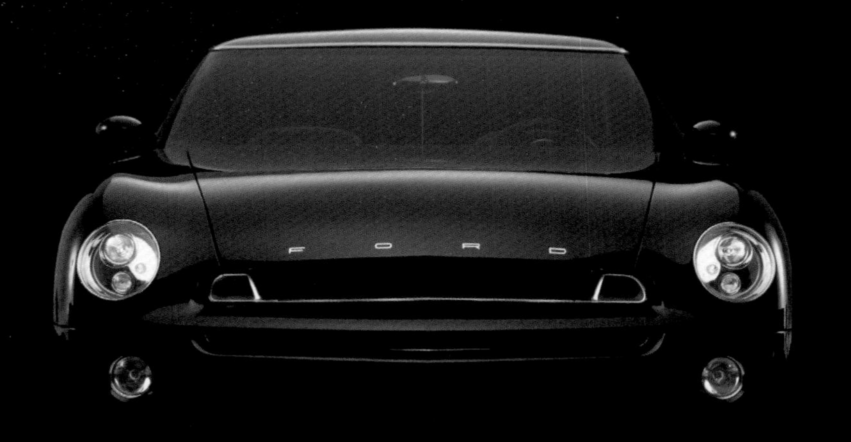

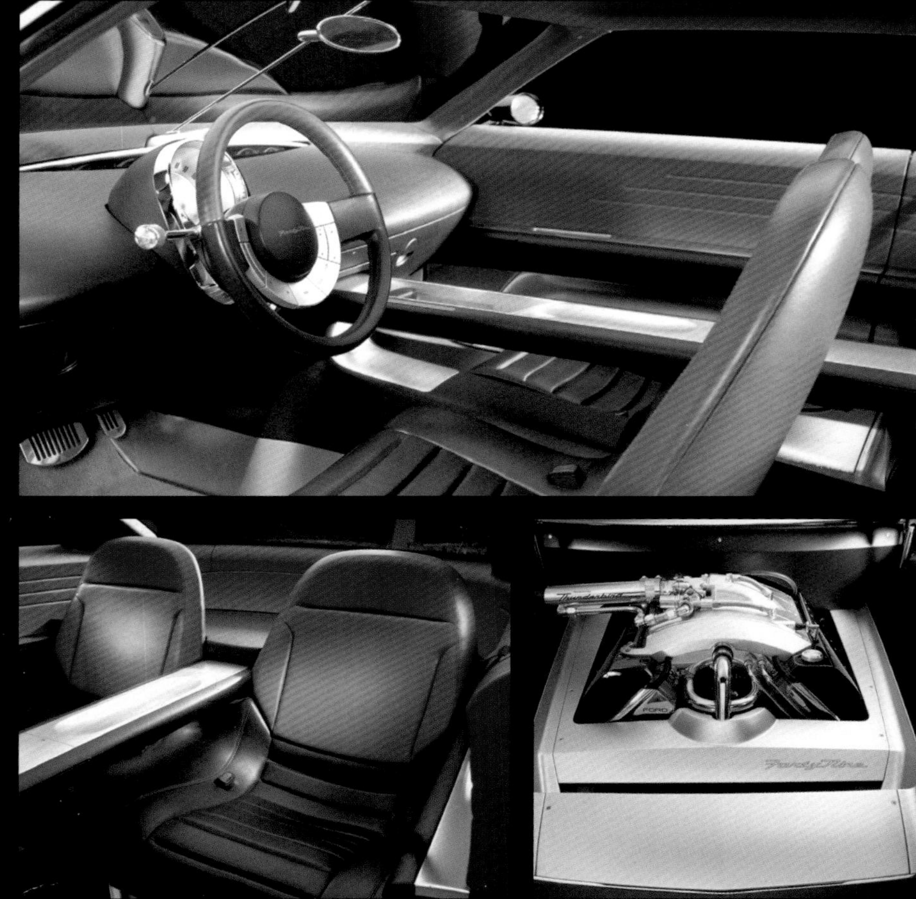

2003

CADILLAC SIXTEEN

The 2003 Cadillac Sixteen recalled the marque's flagship of the Thirties, but the styling and engineering looked to the future. The 1000-horsepower V-16 engine had cylinder deactivation to save fuel and aluminum was used extensively to save weight. The Sixteen was the hit of the auto shows with its dramatic styling and astounding specifications. However, Cadillac decided not to put the Sixteen in production.

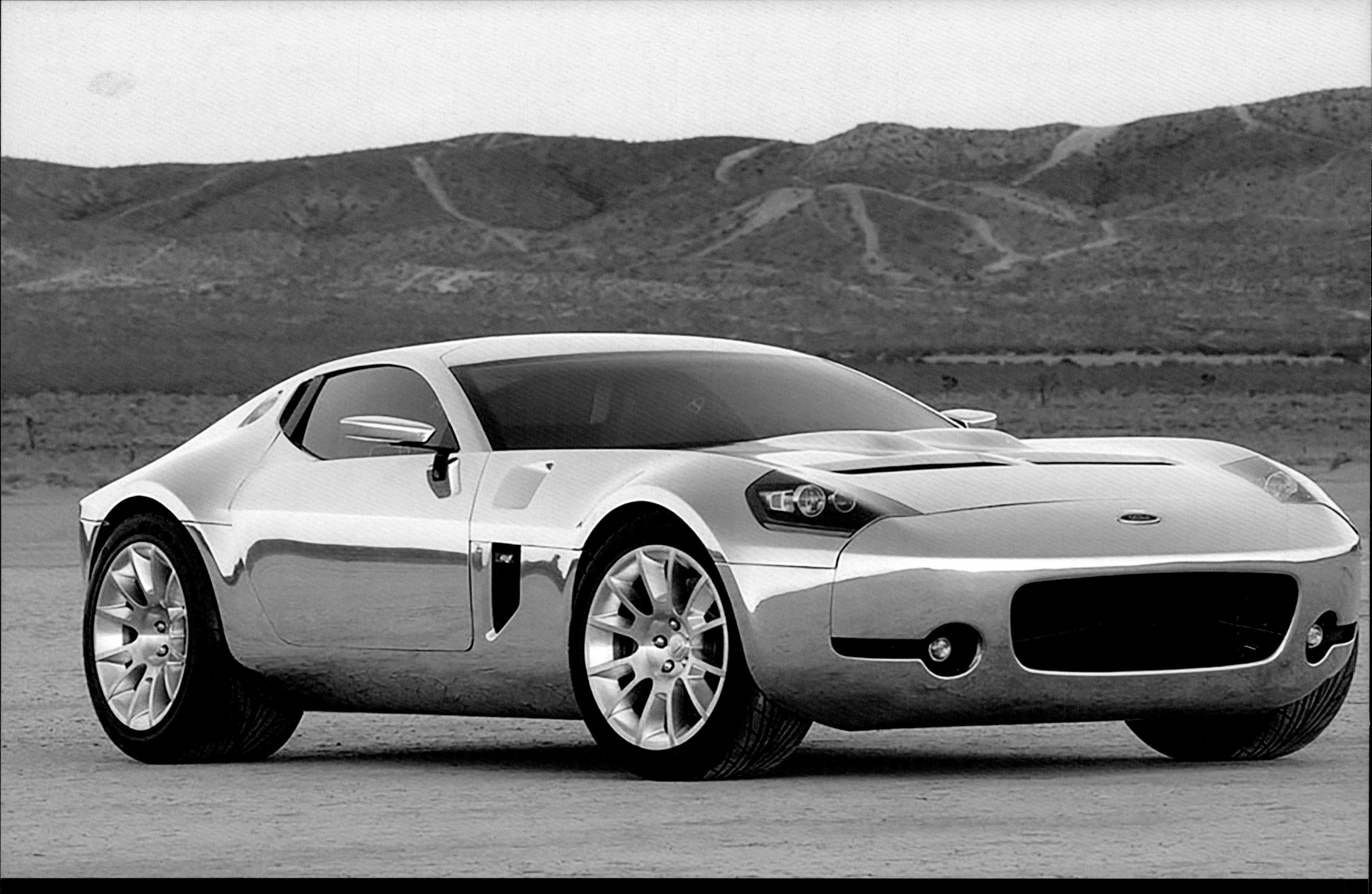

FORD SHELBY GR-1

The 2005 Ford Shelby GR-1 was to be the successor to the 2005-06 Ford GT supercar. The concept car was powered by 605-horsepower 6.4-liter V-10, but a production version would have used a lighter V-8 with similar horsepower. Ford decided not to produce the GR-1 and supercar enthusiasts had to wait until 2017 for a new Ford GT.

BMW MILLE MIGLIA

In 1940, a BMW 328 with an advanced aerodynamic body won the Italian Mille Miglia race. BMW paid tribute to that important victory with the 2006 BMW Concept Coupe Mille Miglia with styling that put a modern spin on the prewar racecar. Under the lightweight body of carbon fiber and aluminum was the drivetrain of a BMW Z4 Coupe.

2006

2007

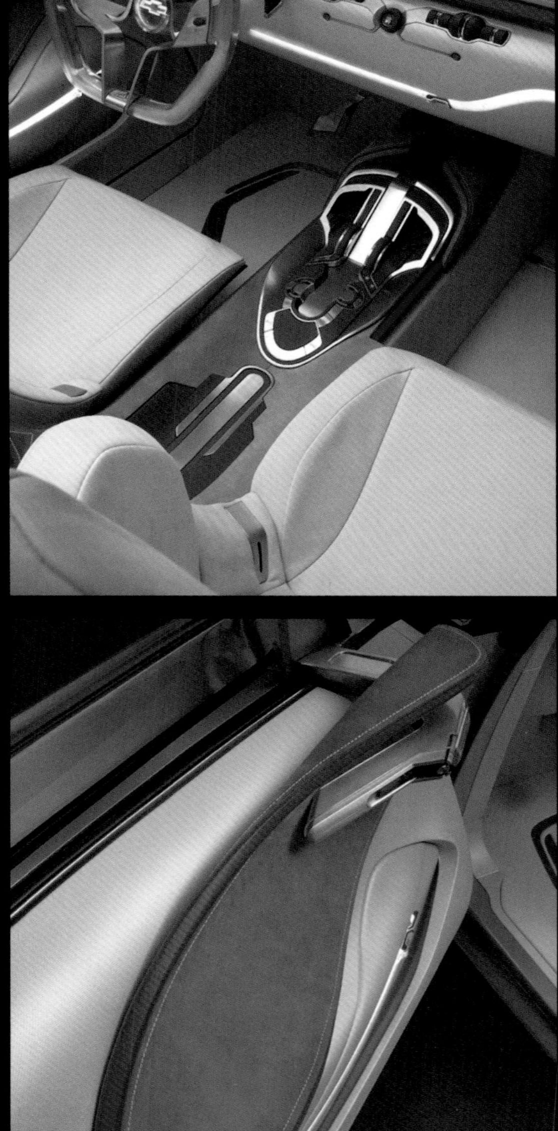

CHEVROLET VOLT

General Motors eased into the electric-car market with the Chevrolet Volt, a car with an electric motor and a "range extender" gasoline engine to generate electricity when the battery ran out of charge. The 2007 Chevrolet Volt concept car was more flamboyant than the more practical production version.

JAGUAR C-X75

The 2011 Jaguar C-X75 supercar concept was a radical hybrid that combined electric motors with a turbine engine. Jaguar announced plans to build 250 C-X75s, but the electric motors would have been combined with a conventional internal-combustion motor instead of the turbine. Projected performance figures were: 0-60 mph in less than three seconds, a top speed in excess of 200 mph, and a 37-mile electric range. Jaguar cancelled the C-X75 after building five prototypes.

2011

BMW 328 HOMMAGE

In 2011, BMW unveiled the BMW 328 Hommage concept car as a tribute to the marque's 328 sports car that debuted 75 years earlier. The original 328 helped establish BMW as a maker of sporting machines and the 328 Hommage harks back to the original with a double kidney-shaped grille and leather hood straps.

CADILLAC CIEL

The 2011 Cadillac Ciel was a large, ultra-luxury, four-door convertible riding on a long 125-inch wheelbase. The Ciel had a hybrid powertrain with a 425-horsepower twin-turbocharged V-6 paired with an electric motor. Cadillac's concept car was unveiled at the Pebble Beach Concours d'Elegance.

CIEL
CADILLAC

2013

ASTON MARTIN CC100 SPEEDSTER

Aston Martin celebrated its 100th anniversary in 2013 with a CC100 Speedster concept car that was inspired by the 1959 Le Mans winning Aston Martin DBR1 racecar. The CC100 packed a 6.0-liter V-12 paired with a six-speed automated-manual transmission. The interior combined carbon fiber and leather.

CADILLAC ELMIRAJ

Cadillac debuted its Elmiraj concept car at the Pebble Beach Concours d'Elegance in 2013. The large coupe didn't enter production, but displayed some future Cadillac style cues. Under the hood was a twin-turbocharged 4.5-liter V-8 with an estimated 500 horsepower.

ELMIRAJ
Cadillac
ELMIRAJ

2015

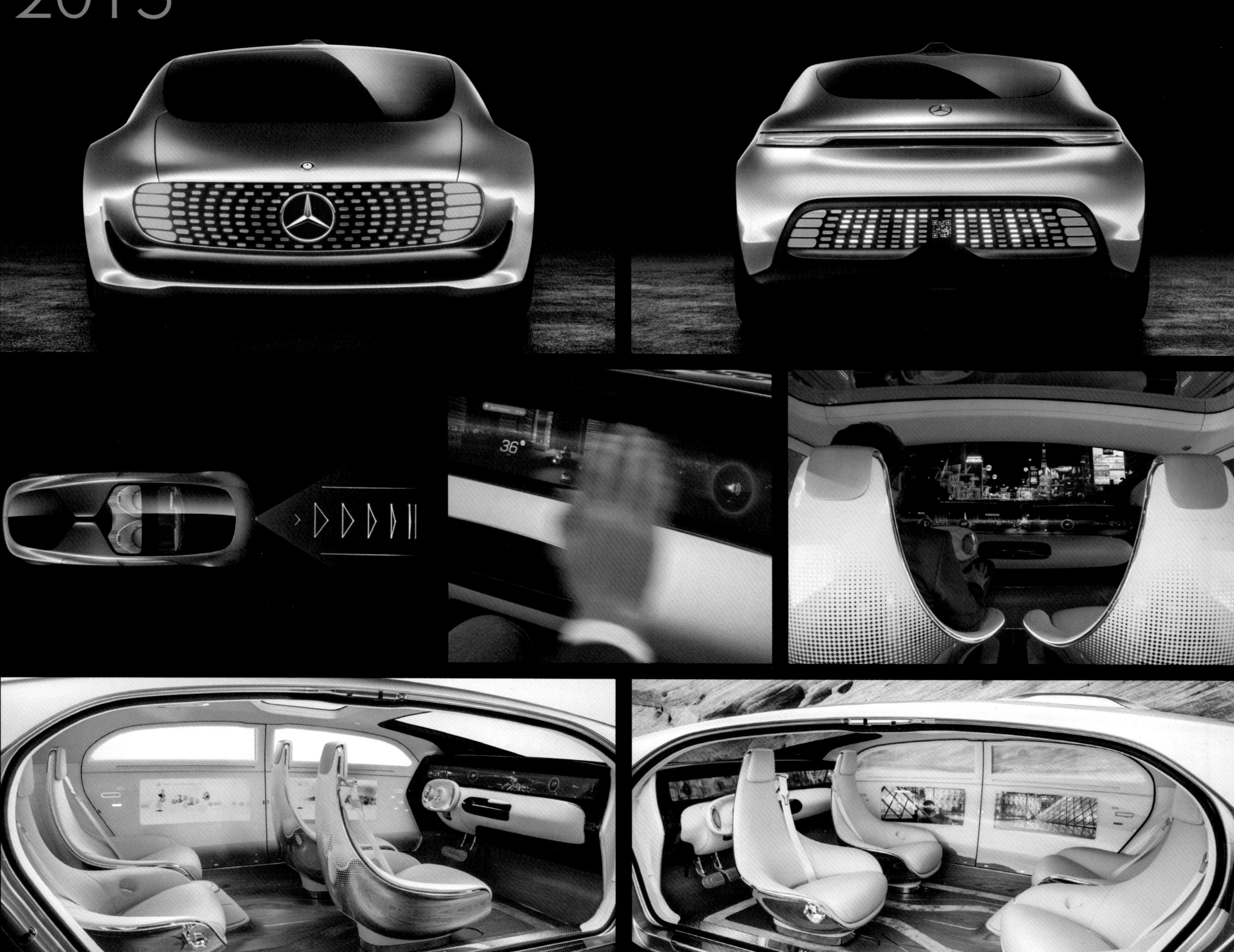

MERCEDES-BENZ F 015 LUXURY IN MOTION

The 2015 Mercedes-Benz F 015 Luxury in Motion concept vehicle was Mercedes' prediction of transportation in the year 2030 when autonomous driving would be commonplace. The Luxury in Motion's front seats could swivel to the rear to aid conversation with the rear passengers. Six screens provided information about the vehicle and a connection with the outside world.

BUICK AVISTA

Coupes were rare on the automotive landscape in 2016 and hardtop coupes (with pillarless side windows) were even less common. The Buick Avista concept hardtop coupe concept shared a platform with the Cadillac ATS and Chevrolet Camaro and seemed feasible as a production car. Although well received, the Avista didn't make it to the showroom floor.

2016

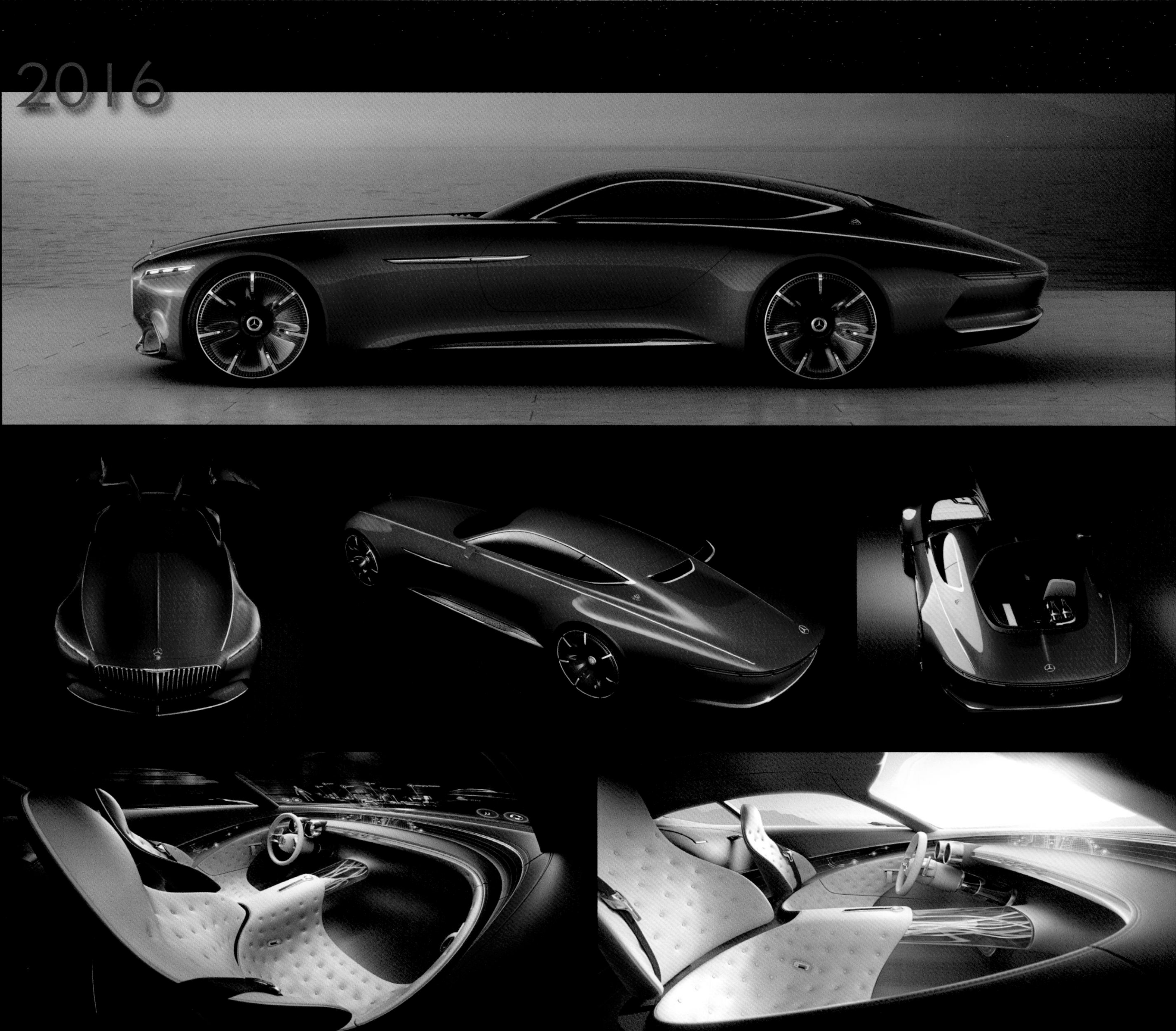
2016

VISION MERCEDES-MAYBACH 6

The 2016 Vision Mercedes-Maybach 6 was a long, sleek coupe concept from Mercedes-Benz's upper-crust niche brand. Although 18.5 feet long, the electric-powered car was said to be capable of 0 to 62 mph in under four seconds and had a range of more than 200 miles.

ROLLS-ROYCE 103EX

The 2016 Rolls-Royce 103EX was the luxury automaker's concept of a driverless car for the year 2040. The vast interior contained a silk couch for only two passengers. An artificial-intelligence assistant bore the name Eleanor, in reference to Eleanor Thornton, the woman said to have been the model for the Rolls-Royce Spirit of Ecstasy mascot.

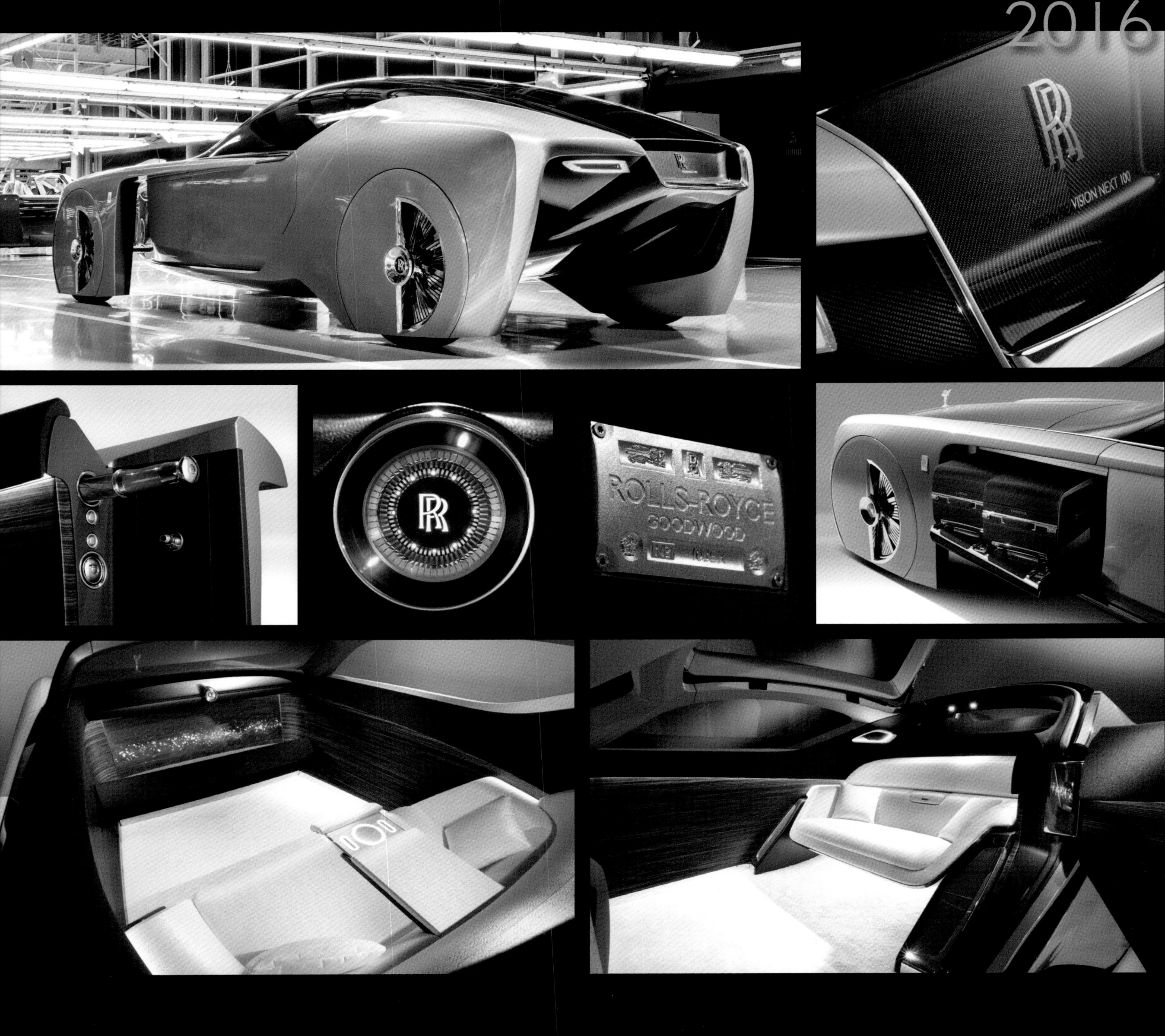
2016
VISION NEXT 100
ROLLS-ROYCE
GOODWOOD

2016
OPENING
30 SUNDAY AUGUST 2015
28 MPH
MENU

VOLKSWAGEN BUDD-E

For many years Volkswagen introduced concepts that suggested that it would build a retro vehicle inspired by its iconic Microbus. In 2016, the BUDD-e concept was an electric version of the Microbus theme. The BUDD-e was also passed over and an electric vehicle with more Microbus-like styling was expected to go into production.

LAMBORGHINI TERZO MILLENNIO

The 2017 Terzo Millennio is Lamborghini's vision for the supercar of the future. Lamborghini teamed up with Massachusetts Institute of Technology to develop a concept for an electric sports car that used a supercapacitor instead of a battery to store energy. An electric motor was incorporated in each wheel to eliminate the need for heavy drive shafts.

2017

2017

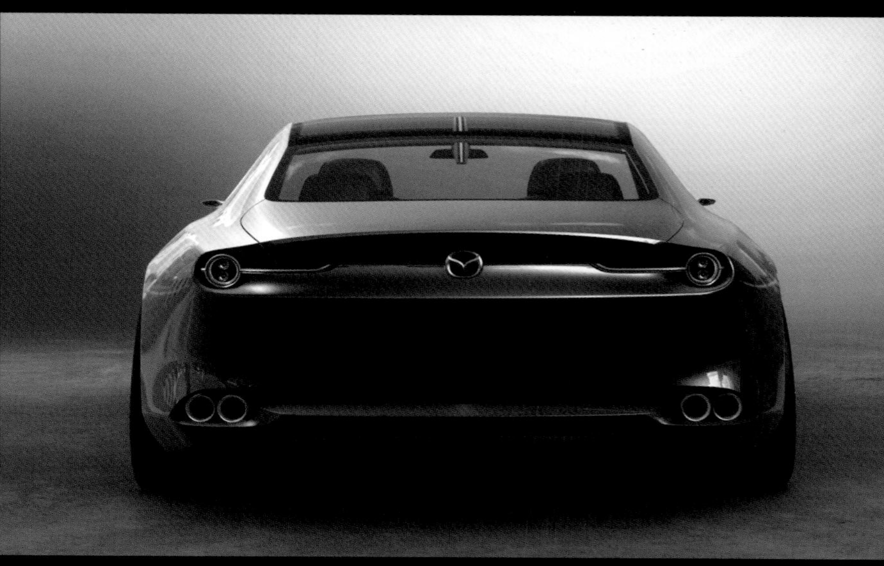

MAZDA VISION COUPE

Automakers often call a sleek four-door sedan a "coupe" because of the car's smooth, coupe-like roofline. A case in point is the 2017 Mazda Vision Coupe. The Vision Coupe is a continuation of the design theme of the red 2015 Mazda RX-Vision concept—a car that was a true coupe.

NISSAN VMOTION 2.0

The 2017 Vmotion 2.0 concept car suggested the direction of Nissan design for future sedans. It also showcased the latest version of Nissan's ProPilot automated driving system that could drive the car within a selected lane. When ProPilot was operating, badges on the front and rear of the car glowed.

2017
NISSAN
Vmotion 2.0

2017

TESLA ROADSTER

Before Tesla built sleek electric sedans, the carmaker launched with a small electric sports car. The first Tesla Roadster went out of production in 2011 and the concept car displayed in 2017 pointed the way for a second-generation Roadster. Tesla claimed the new Roadster would accelerate 0 to 60 mph in 1.9 seconds and have a top speed of 250 mph.

HYUNDAI LE FIL ROUGE

The 2018 Le Fil Rouge concept car introduced Hyundai's "Sensuous Sportiness" design theme that was expected to influence future Hyundai models. The interior used revitalized wood and high-tech fabrics.

2018

2018

JEEP 4SPEED

Jeep typically introduced modified versions of its products at its Easter Jeep Safari in Moab, Utah. For 2018, it showed a special 4SPEED that was 22 inches shorter and 950 pounds lighter than the production Wrangler. Those changes made the Jeep even more agile off road and, of course, faster.

LAGONDA VISION CONCEPT

Aston Martin bought the Lagonda make in 1947 and often used the Lagonda name on its sedan models. The Lagonda Vision Concept was a vision of the self-driving, electric car of the future. With the battery mounted under the floor and no need for a bulky gas engine, the Vision Concept devoted most of its space to the passenger compartment.

2018

BMW
328 Hommage